CLOSE-QUARTER COMBAT

This book is dedicated to the men and women of the U. S. Armed Forces law enforcement community and all who seek a realistic approach to unarmed self-defense.

CLOSE-QUARTER COMBAT

A Soldier's Guide to Hand-to-Hand Fighting

Professor
Leonard Holifield

Paladin Press
Boulder, Colorado

Close-Quarter Combat: A Soldier's Guide to Hand-to-Hand Fighting
by Professor Leonard Holifield

ISBN 0-87364-924-9
Printed in the United States of America

Published by Paladin Press, a division of
Paladin Enterprises, Inc.
Gunbarrel Tech Center
7077 Winchester Circle
Boulder, Colorado 80301 USA
+1.303.443.7250

Direct inquiries and/or orders to the above address.

Contents

Warning

Some of the techniques and drills depicted in this book are extremely dangerous. It is not the intent of the author, publisher, or distributors of this book to encourage readers to attempt any of these techniques or drills without proper professional supervision and training. Attempting to do so can result in severe injury or death. Do not attempt any of these techniques or drills without the supervision of a certified instructor.

The author, publisher, and distributors of this book disclaim any liability from any damage or injuries of any type that a reader or user of information contained in this book may encounter from the use of said information. This book is presented for *academic study only.*

Disclaimer

Please note that the publisher of this instructional book is NOT RESPONSIBLE in any way for any injury that may occur as a result of reading or following the instructions herein. This book was written as a technical guide for qualified military and law enforcement personnel, with no intention of contradicting their current department or agency policies. It is essential that before following any of the activities described, physical or otherwise, readers consult their physicians, since some of these activities are complex.

Acknowledgments

This book would not have been possible without the support and assistance of the following individuals:

Ron Simmons and David Howell for their outstanding photography. Elena Ontiveros for her fine graphics and technical expertise. Michael Green, Jim Bates, Alan Cohen, Cecil Nelson, Kyle Byram, Aaron Spaulding, Jason Higgins, and Stephen Chandler for their assistance in the demonstration of the techniques presented. And Bob Newman for his assistance and technical expertise in the final production of this book. Special thanks go to my loving parents, who brought me into this world and taught me the value of dedication and hard work; my loving wife and best friend, Ena, for believing in me and my dreams; and to the individuals listed below, who, through their support and guidance, helped me to bring the importance of this training to the forefront:

LTC Dennis Lewis
LTC Mark Rambis
LTC Arnaldo Claudio
SMA Richard A. Kidd (Ret)
CSM William Perry (Ret)
CSM Leslie Freeman Jr.
CSM Stanley Mitchell (Ret)
1SG Marlin Esh
Alice Edwards
Greg Willis
Bill Hardy
Kay Mundy

Very special thanks go to Professor E.A. Moore, president of the World Martial Arts Hall of Fame, and to my assistant instructor staff, who through the years have dedicated themselves to the mission of providing the best training possible to the soldiers of the U.S. Armed Forces:

James E. Turner, Maceo Jourdan, Donald Mouton, Michael Maxwell, Irving Domenech, Hence Spencer, Andrew Hewuse, Christopher Harrison, Michael Murkins, Richard Bush, James Gressley, Michael Pennington, Thomas Felder, Lidel Cordero, Butch Nunley, Jason Higgins, David Howell, and Stephen Chandler.

The destination is not as important as is the process of getting there.

—Professor Leonard Holifield

CHAPTER ONE

The Military and the Martial Arts

To clearly see the close relationship between the military and the martial arts, let's first define the two. Webster's Dictionary defines *martial* as "relating to or suited for war or a warrior." The word *military* is defined as "pertaining to or characteristic of soldiers or the armed forces, or pertaining to war." The two definitions are very similar and they share one major characteristic that is vital to the successful accomplishment of any mission, be it in the dojo or on the battlefield: discipline.

From the first day of basic training a soldier learns that discipline and following orders are essential if one is to survive on the battlefield. The same applies to the new student entering the martial arts dojo to study his chosen art. The student quickly learns that martial courtesy, respect, ethics, and discipline are required in order to learn and perfect the skills needed to obtain the coveted black belt.

Discipline is not the only major bond between the military and martial arts; both teach the principles of defense, be it in the form of self-defense or the defense of one's country. On the battlefield the soldier must be physically fit and able to think clearly, make quick and timely decisions, and carry out his plans in such a manner so as to disrupt the enemy and accomplish his mission. Likewise, the martial artist must also be in top physical condition and be able to think and react in a manner that will ensure victory in the tournament ring or survival in a street attack.

Now that you have a better understanding of the relationship between the military and the martial arts, let's think about discipline. Discipline can be defined as training that is expected to produce a specific character or pattern of behavior, especially training that produces moral or mental improvement. The martial arts are based on sound martial ethics designed to build good character and morals in an individual for the purpose of living and maintaining an honest and healthy life. The military has a long and distinguished history of producing well-disciplined soldiers who have fought in wars past in the defense of their country. Mental conditioning is the direct result of individual discipline, and any discipline a soldier learns can only enhance the discipline required on the battlefield.

The Natural Weapons of the Body

The human body is more complex than the most advanced computer technology. So complex and self-contained is it that it has its own internal immune system designed to defend against attack by unfriendly organisms. It must also have a means of protection to defend against a violent attack from a dangerous foe.

Your body is well equipped with a variety of weapons to aid in its defense. As a hand-to-hand combat specialist, you must be able to use these natural weapons effectively in the event that you find yourself in a close-quarter combat situation. Your body's natural weapons include hands, forearms, elbows, fingers, feet, legs, knees, head, and teeth. By learning to use these natural weapons you can direct effective strikes to the weak areas of your attacker's body, immobilizing, incapacitating, or killing him if necessary. The following photographs describe each of the natural weapons and areas you should target for maximum results.

THE FIST

Target Points

The jaw, nose, eyes, throat, sides of the neck, solar plexus, ribs, kidneys, and groin.

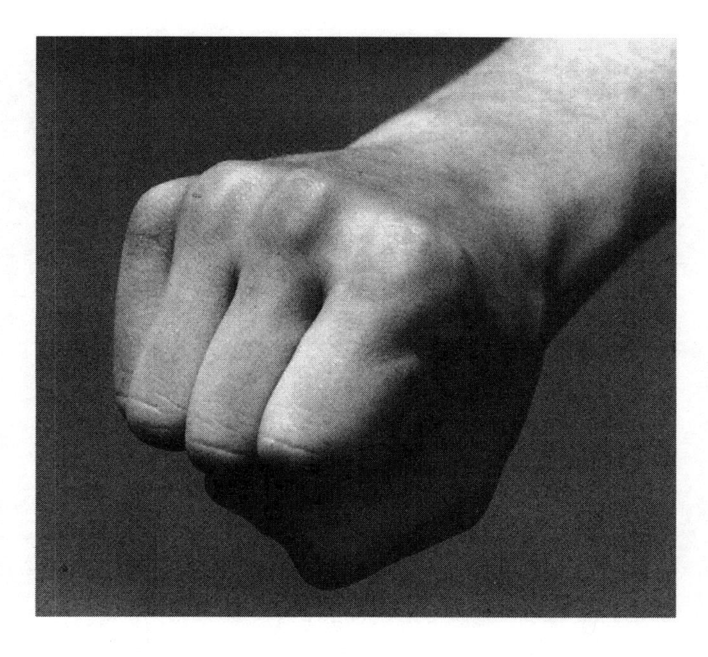

Your fists will serve as your primary weapons in most close-quarter engagements. The hand, when formed into a tight fist, can be a formidable weapon. The point of contact is at the first two knuckles or the outside portion of the fist near the baby finger (hammer strike).

THE FOREARM

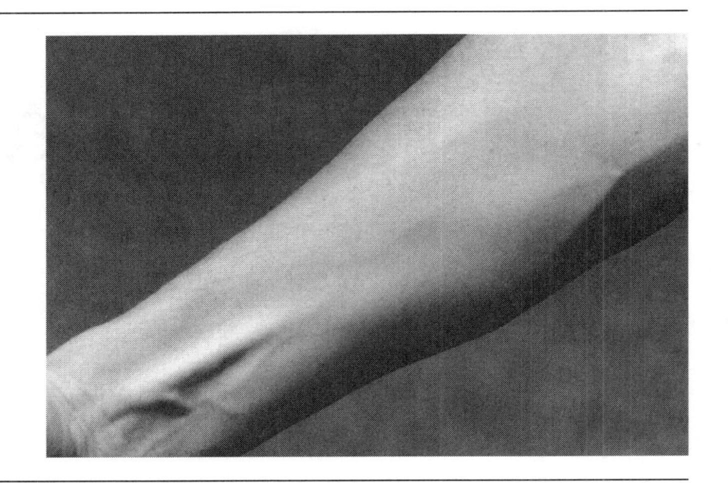

Target Points

The throat, side and back of neck, and groin.

Your forearm can be used as a devastating weapon because of its bone strength and the power that can be generated by the arm. The striking areas include the outside and inside edges of the forearm.

THE ELBOW

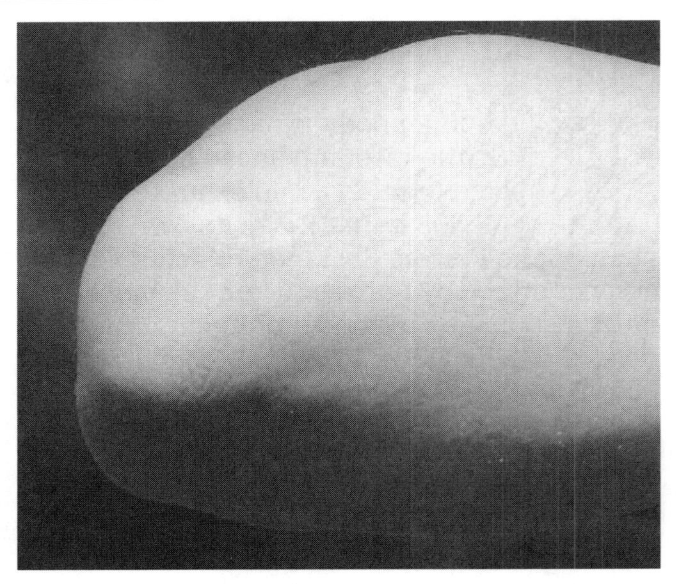

Target Points

The face, sides of head, sides of neck, throat, ribs, solar plexus, kidneys, groin, and knees.

Your elbow is a powerful weapon that can be delivered at four different angles. The striking angles of the elbow include the forward, reverse, upward, and downward strikes.

THE FOOT

Target Points

The solar plexus, ribs, kidneys, spine, coccyx, groin, thighs, knees, shins, and insteps.

Your foot can be used to deliver a variety of effective and crippling kicks. The striking areas of the foot include the ball, instep, outer edge, inner edge, heel, and bottom.

THE KNEE

Target Points

The head, solar plexus, groin, kidneys, spine, and outer thighs.

Your knee can be used effectively on several points of the body.

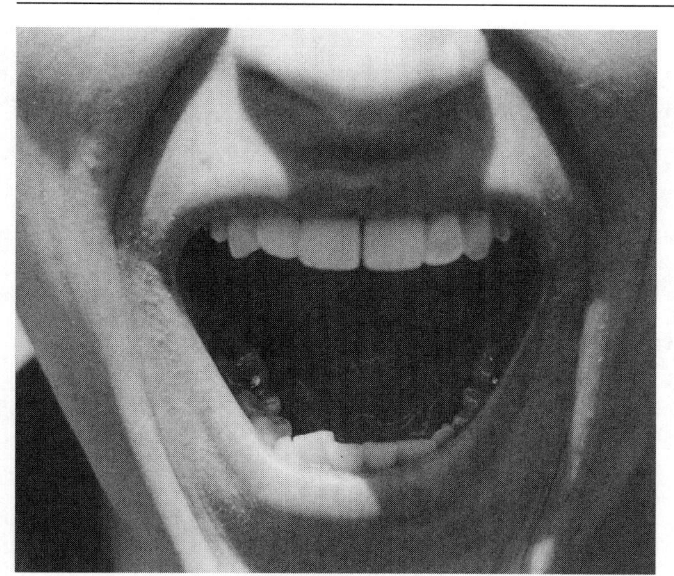

THE TEETH

Target Points

The ears, nose, neck, throat, and groin.

Your jaw is capable of applying tremendous biting pressure. This alone makes the teeth an extremely effective weapon.

THE HEAD

Target Points

The head and face, solar plexus (ramming head butt), groin, kidneys, and knees.

Your head can also be used as a weapon by executing a forward head butt to the attacker's face or forehead. Caution must be exercised here to avoid injuring yourself; try to strike with the hairline area, which is the strongest point.

THE LEGS

Target Points

The head, groin, neck, solar plexus, kidneys, knees, arms, and ankles, to name a few.

Your legs are your most powerful weapons and can be used against a variety of targets by using kicks, sweeps, and stomps.

EXTENDED FINGERS

Target Points

The eyes, throat notch, underside of the jawbone, area behind the ears, and the nostrils.

Your fingers can be used to jab into or apply direct pressure to soft tissue targets.

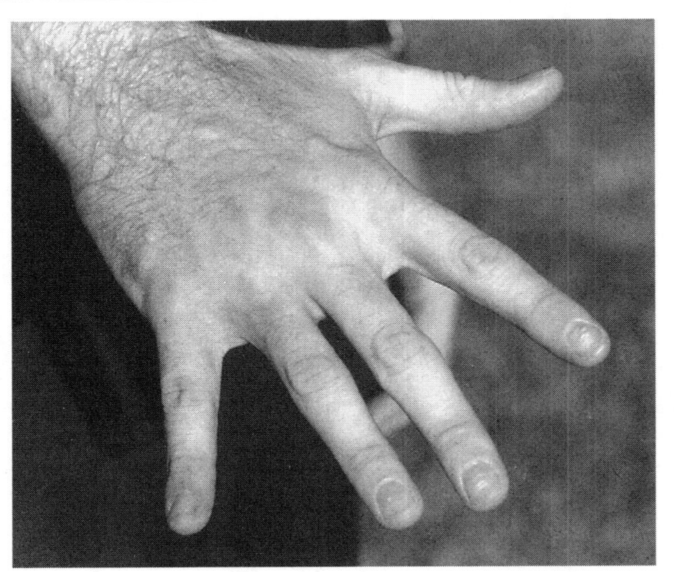

Vital Points of the Body

To be an effective fighter one must have a clear understanding of the human anatomy—more specifically, the points on the body that you can strike to debilitate your attacker immediately. This chapter will cover the most effective striking points used for close-quarter combat. (These striking points can cause immediate incapacitation, crippling, or death, which means that you should take extreme care when practicing with your partner.) An effective fighter will also avoid striking vital target areas that may be protected by combat gear such as a Kevlar helmet or flak vest. Doing so would only waste the time and energy you need to quickly take out your attacker.

PRESSURE POINTS

Pressure points are those points on the body that, when manipulated, will cause mild to extreme pain. Applyimg pressure to these points is most effective during a joint-locking technique when the attacker must be controlled but not injured or killed. Use of these points is especially well-suited for law enforcement personnel when allegations of police brutality or excessive force may be a concern.

Underneath the Nose

Apply finger pressure applied directly under the nose is an effective way to get an attacker off of you. A punch directly to the nose can break it and cause the eyes to water. A palm-heel strike directed under the nose with force can be a lethal strike when crushed nasal bones cause internal hemorrhaging in the frontal lobes of the brain, not because of bones shooting directly into the brain (as goes the common misconception).

The Hollow of the Throat

Another pain point is just below the Adam's apple at the hollow of the throat and appears as a small notch. Pressure to this point should be applied inward and downward for maximum effectiveness. A forceful strike made directly into this area can cause death by collapsing the windpipe.

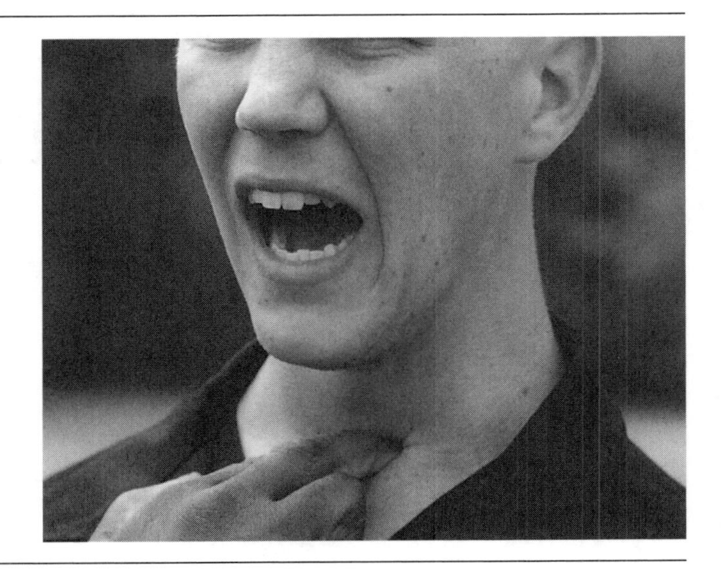

The Trachea

The trachea is a vital target that, if crushed or stricken with sufficient force, can cause death by destroying the windpipe a little above the hollow of the throat. The trachea is an excellent target of opportunity.

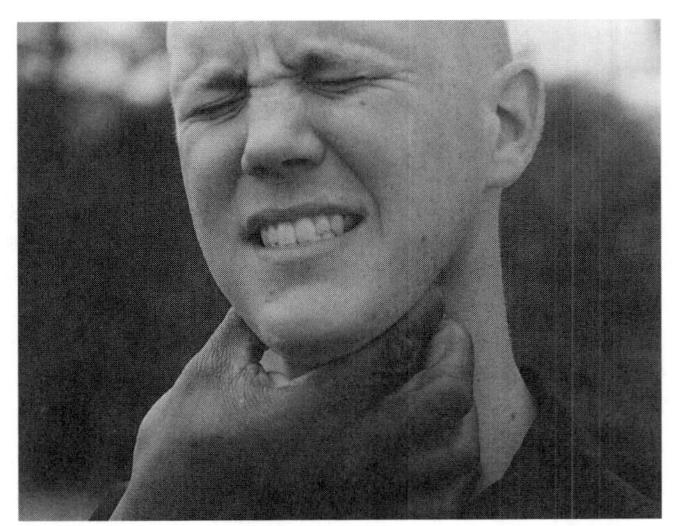

The Area Behind the Ears

Direct pressure applied behind the ears will cause enough pain (mild to extreme) to distract the attacker long enough to allow you to follow up with your primary technique.

The Clavicle Notch

This is an extremely painful point when pressure is applied directly behind the clavicle (collarbone) in a downward motion.

The Inside Elbow Notch

This point can be used to disarm or control your opponent. Pressure is applied directly into the notch.

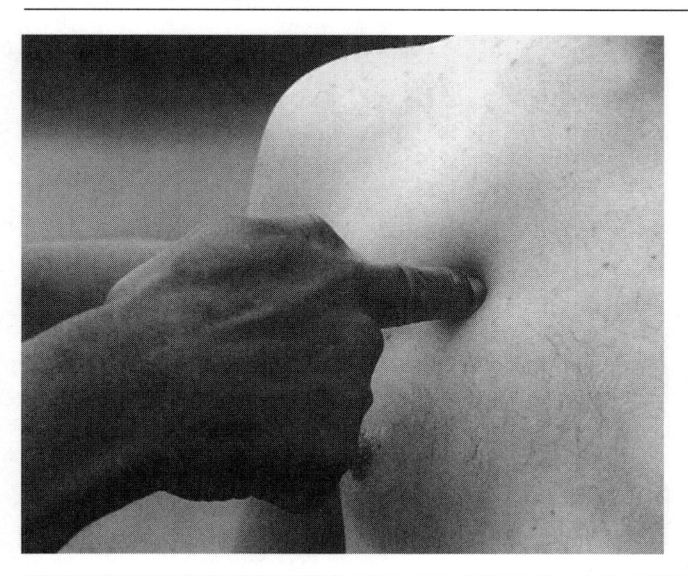

The Pectoral Muscle

When pressure is applied directly into the pectoral muscle, it is an excellent controlling tactic because it causes extreme pain.

The Edge of the Pectoral Pinch

By grabbing the pectoral muscle at the edge and pinching a good chunk of skin and muscle, extreme pain can be achieved.

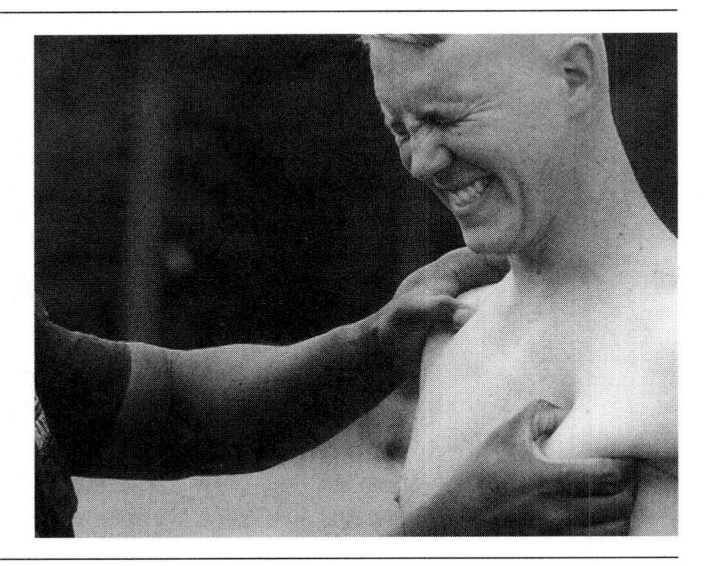

The Solar Plexus

The solar plexus is located at the point where your rib cage meets at the center of your chest. It is a very tender area that is also very vulnerable to a well-placed strike, which can easily knock the wind out of an opponent.

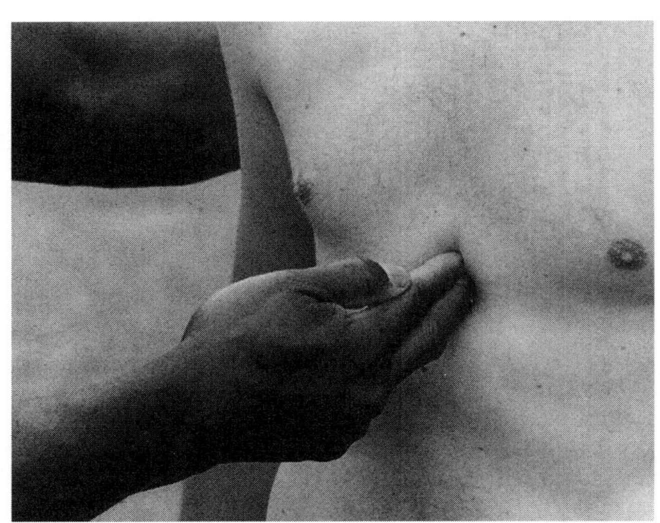

The Obliques

The obliques, or "love handles," as they are commonly known, can be grabbed and squeezed, causing intense pain.

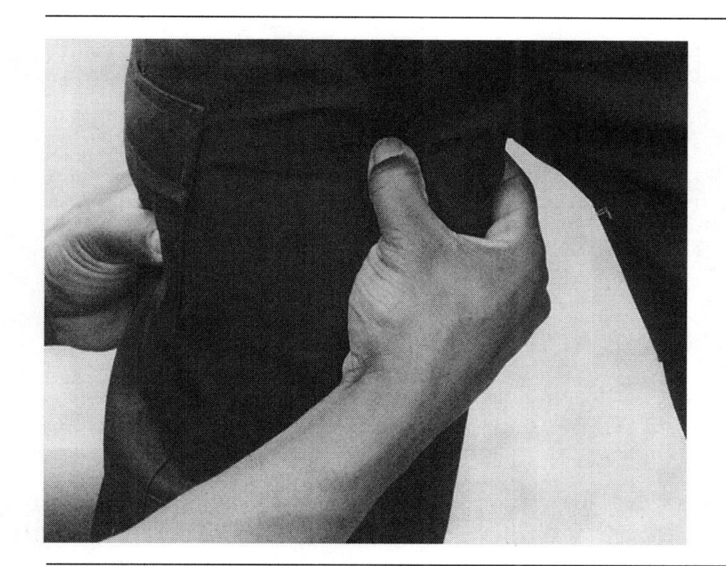

The Inside of the Thigh

This point is approximately two inches down from where the top of the thigh and bottom of the trunk meet on the inside. As you pinch deeply into this area, you will feel the muscle and tendons. This pinch is an excellent tactic to use against a headlock.

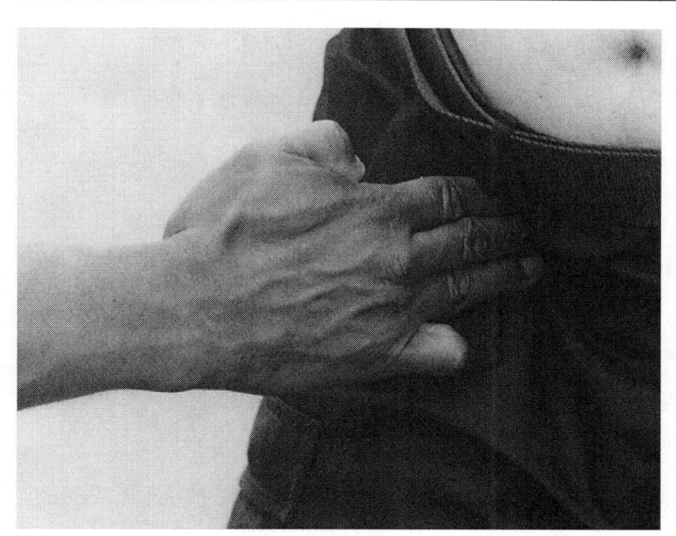

The Pelvic Triangle

A strike or kick directed at the point of the pelvic triangle can easily disable an attacker's leg.

Physical Training

This chapter will cover specific exercises geared toward developing strength, stamina, and flexibility. Because of the added stress such training places on the joints, proper joint preparation exercises and stretches will be addressed. You should perform these before beginning your training session, and you can do so alone or with a partner.

JOINT PREPARATION

NOTE: All stretches in this section should be done clockwise and then counterclockwise, if possible. Each stretch is repeated until that muscle group feels loose and relaxed.

Neck Rotations

Rotate the neck slowly in full circles to loosen and relax it.

Shoulder Rotations

Rotate both shoulders in a circular motion to loosen the muscles and joints.

Hip Rotations

With hands on your hips, rotate your hips in a full circular motion.

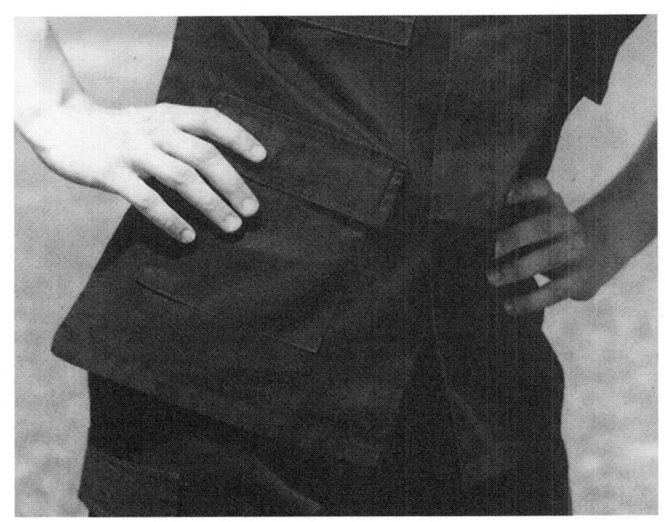

Knee Rotations

Place both feet together, bend slightly at the knees, and rotate the knees.

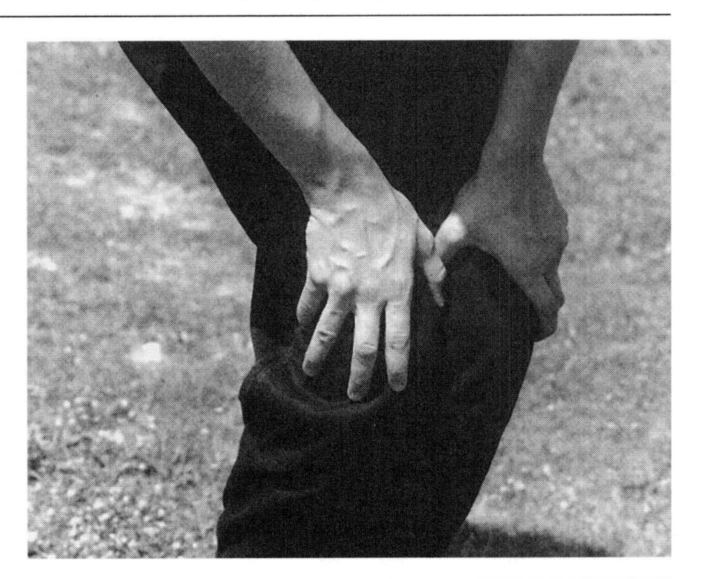

Ankle Rotations

These can be done either standing or sitting. From the standing position, place one foot behind you with the toes pointed toward the ground and rotate the ankle, then switch feet. From the seated position, sit with your legs folded in front of you as you grab your ankle and rotate. Switch feet.

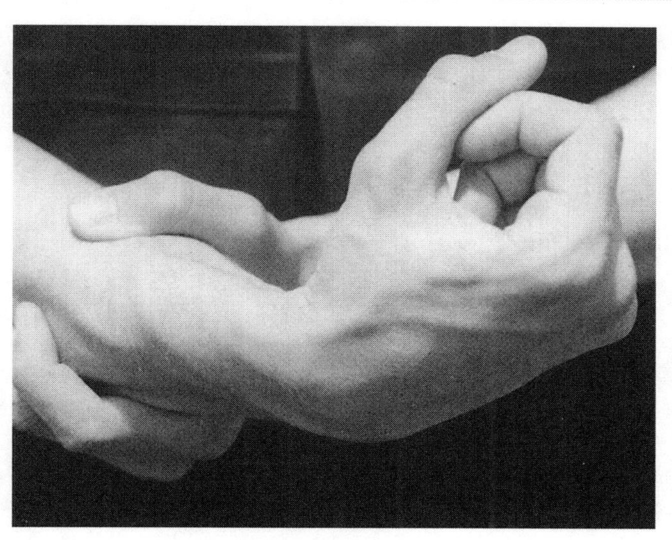

Wrist Rotations

Rotate the wrists in full circles. The wrists must be warmed up properly prior to performing joint manipulation techniques that use the wrists.

STRETCHING

Neck Stretch

While standing, stretch the neck by turning the head slowly to each side as far as possible, then forward and backward.

Chest Stretch

From the standing position, place your arms straight out in front of you with your hands together. Spread your arms wide apart to stretch the chest muscles and then return to the standing position.

One-Arm Side Stretch

While standing, bring one arm up and over your head, stretching that side, then switch arms.

Backroll Stretch

Lie flat on your back with your arms at your side. Roll backwards, bringing both legs up and over your head as far as possible but without going completely over. Hold the position and then bring your legs down slowly.

Leg Spreader

From the seated position, spread your legs as far as possible. Stretch forward toward the center and then slowly pivot to each foot. Return to the starting position and repeat.

Hamstring Stretch

While standing, bend forward with your legs as straight as possible until your head is at about knee level. Your hands should be grasping the lower leg. Hold there until the burning stops and then slowly stand erect.

Groin Stretch

While seated, bring both legs in toward the body with the soles of your feet touching. Stretch by applying outside pressure to the inside of the knees with both elbows.

Long Sit Stretch

Still seated, place both legs together and lean forward; attempt to touch your toes simultaneously with both hands.

Elongation (Supine) Stretch

Lying flat on your back, reach straight back behind and over your head while pointing your toes, stretching your entire body.

Seated Side Stretch

Spread your legs as wide as possible and reach over one leg with both arms; attempt to touch the ground with your nose. Now switch to the other leg.

UPPER-BODY STRENGTHENING

Regular Push-Ups

With your arms about shoulder width apart, lower your rigid body to a point approximately one or two inches off the ground and push back up. Do not arch your back; try to keep your body straight and rigid like a board.

Wide-Arm Push-Ups

To work the outer pectoral muscles, spread your arms beyond shoulder width in the push-up position. The exercise is performed the same as a regular push-up.

Diamond (Close Hand) Push-Ups

The inner pectoral muscles and triceps can be strengthened by doing push-ups with the hands placed together forming a diamond shape between them.

4-Level Elevated Push-Ups

With your partner holding your ankles off the ground at knee level, do a push-up. Next, have your partner raise your ankles to waist level, while you do another push-up. Repeat at chest level, and finally shoulder level. Then your partner takes you progressively back down and you start over again. When at shoulder level, you should place your insteps on your partner's shoulders while he crosses his arms over the back of your ankles to secure your legs. This type of push-up works the upper shoulders, back, and biceps.

4-Man Chain Push-Ups

These are done by having four men lie face down at 90-degree angles to each other, with the feet of one man placed over the lower back of another. The team members must work together to perform the exercise at the same time.

MID-SECTION ABDOMINAL STRENGTHENING

Regular Sit-Ups

Seated with knees bent and your hands behind your head, exhale while raising your upper torso until your nose breaks the imaginary plane between your knees, then inhale as you go back down slowly (using abdominal and upper leg muscles coming up and going down).

Twisting Sit-Ups

This exercise is done the same as a regular sit-up, but you twist to one side and then the other as you reach the knees, touching your right elbow to the left knee and vice versa.

NOTE: Regular (full) sit-ups and twisting sit-ups are, according to some fitness instructors and medical professionals, potentially injurious to the lower back. However, many people do them and experience no pain or injury. You make the call.

Crunches

Lying on your back with your knees bent and arms across your chest, raise your torso to approximately 45 degrees, but no more. (This places total emphasis on the abdominal muscles.) Hold this position for a moment and return to the starting position, then repeat. Exhale on your way up and during the hold. (Crunches have largely replaced the sit-up because of the aforementioned concern over injuring the lower back.) Note that you can also do these to the side to work the obliques.

Leg and Head Lifts

Lying on your back with your hands beneath your lower back where it meets the upper buttocks and your legs straight, raise your head and feet approximately six inches off the ground. Hold this position for a moment and then go back down. Repeat.

CHAPTER FIVE

Stances: The Foundation of Balance

One of the most important elements in hand-to-hand combat training is a strong, well-placed stance. From the stance you will launch your attack or execute your defense. Your stance must be stable and your body weight evenly distributed in order to maintain balance upon contact with your attacker. In general, the stance that you use will often depend upon the type of attack being directed at you. A common mistake is to predetermine your defense before the attacker strikes. This can cause you to go into an ineffective stance that can result in loss of balance and possibly defeat. You can correct this mistake by shadowing, or making yourself one with, your attacker. By doing this you flow with your attacker and whatever he throws at you; the stance you need at any given time will happen naturally.

This chapter will cover the two most common stances used in close-quarter combat training, with emphasis on the fighting stance. These two stances provide the stability and balance needed to meet a violent attack. Other traditional karate stances, such as the cat stance, do not offer the stability required in close-quarter combat.

THE FIGHTING STANCE

Standing with your legs equally spread more than shoulder distance apart, center your weight and adjust your equilibrium so that you are well balanced. Make a relaxed fist with your hands and place them up and in front of your body for protection. A tightly closed fist can add unwanted tension and take away from the fluid and relaxed movement required for executing strikes. The fist will naturally close tightly before actual impact.

It is best to move on the balls of your feet rather than flat-footed, which also can reduce fluidity of motion.

THE BACK STANCE

From the fighting stance, shift your body slightly to the rear, putting a little more distance between you and your attacker.

NOTE: Do not lock the lead leg at the knee; keep it slightly bent to avoid hyperextension in the event that it is kicked.

Blocking Drills

This chapter will cover the four basic blocks used in close-quarter combat: the high, low, outside, and inside blocks. Variations of these can be found in most traditional martial arts systems. The difference here is how each block is executed, which is from a relaxed fighting position, with the hands slightly open for ease and speed of movement.

HIGH BLOCK

The high block is used to defend against overhead attacks. The blocking arm should be at a 45-degree angle to allow for deflection of the attacking arm.

LOW BLOCK

The low block is primarily used to defend against low-line kicks. When executing this block, be sure that your lead leg is bent (not locked) to protect it from being hyperextended in case it is kicked.

OUTSIDE BLOCK

This block is used to defend against hook or straight punches. You must move your body to the left or the right (getting it out of the line of attack) while executing this block. This will also allow you better target selection for your counterstrike.

INSIDE BLOCK (CHECK)

This is an extremely effective block when executed correctly. As the attacker punches, inside-block his hand with your right hand while bringing your left hand, palm facing outward, to the side of your face. This position allows you to block the attacker's punch while defending your face against a secondary attack.

INSIDE BLOCK (TWO)

Quickly check the punching hand by grabbing it with your left hand and slipping your right hand under your armpit. From this position you can easily pull the attacker into a backhand strike.

Kicking Drills

All kicks presented in this chapter are mid- to low-line level kicks, directed no higher than the solar plexus. Kicking higher than the solar plexus can place you in an off-balanced position. The objective of kicks used in close-quarter combat fighting is to destroy your attacker's foundation by attacking his legs, knees, shins, groin, and so on. High, flashy kicks, including spinning kicks, are not advised because they can leave you off balance and vulnerable to a counterattack.

The following kicks are recommended and should be practiced using extreme caution to prevent injury to yourself, your training partner, and your trainees.

FRONT THRUST KICK

From the fighting stance

raise your rear leg

and thrust the kick out toward the target.

INVERTED FRONT KICK

This kick works best at close range. From the fighting stance bring the rear leg up and out toward the target (notice the angle of the foot). Contact is made with the bottom of the foot. Note that this kick works best at close range.

SIDE THRUST KICK

From the fighting stance bring your rear leg up into your chest

and thrust the kick out (again notice the angle of the foot) until contact is made with the blade of the foot.

LOW-LINE ROUND KICK

From the fighting stance bring your rear leg up

and execute the round kick in a hooking motion toward the target. This is an excellent kick for attacking the knee.

THE STOMP

A strong finishing technique, the stomp can be directed against the face, head, throat, sides of neck, groin, ribs, solar plexus, knees, and ankles.

Striking Drills

The strikes covered in this chapter are geared toward inflicting maximum damage to the vital points and internal organs of the body. Remember, in combat there are no rules.

Control should be exercised at all times when practicing these techniques with your partner. Emphasis should be placed on correct movement (maintaining balance), precision, power, and speed.

STRAIGHT PUNCH

From the fighting stance, punch straight out to the target, making contact with the first two knuckles of the striking hand.

BACKHAND JAB

From the fighting stance, jab the lead hand out toward the target making contact with the knuckles of the back of the fist; quickly rechamber the hand for the next strike.

HOOK PUNCH

The hook punch can be used effectively at close quarters. From the fighting stance, strike in a hooking motion to the jaw, side of the neck, ribs, groin, chin, or kidneys.

UPPERCUT

The uppercut, like the hook punch, can be quite effective at close quarters to attack the solar plexus, midsection, or chin.

OVERHEAD PUNCH

From midrange, bring the punch over the top in an arc toward the attacker's face.

HAMMER STRIKE

An extremely powerful strike, the hammer strike is used against the head, neck, back, ribs, kidneys, heart, and knees. Strike downward toward the target, making contact with the bottom, meaty portion of your fist.

ELBOW STRIKE

From the fighting stance, the elbow strike can be executed from a stationary position or while moving in on your attacker. It can be executed from four angles of attack: forward, reverse, upward, and downward.

PALM-HEEL STRIKE

From the fighting stance, execute the palm-heel strike toward the attacker's chin, snapping the head back. This is a very effective strike, and it can also be directed to the knees, ribs, solar plexus, and head.

SIDE-THRUST PUNCH

When executing this strike, turn your shoulder into the strike; this can add up to six inches of penetration to the punch.

V-STRIKE

The V-strike is used to attack the throat or underneath the nose.

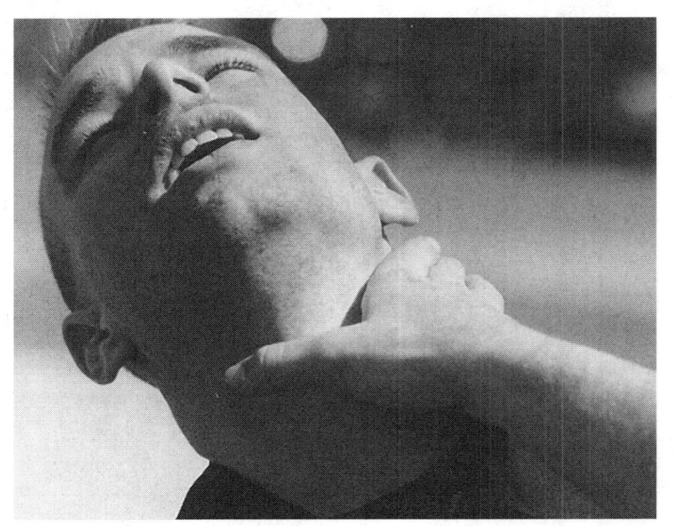

Combat Psychology and Mental Conditiong

The final outcome of any physical confrontation will depend solely on your state of mind. You must develop a mind-set that will allow you to engage your attacker without hesitation or fear of injury. Fear can be your worst enemy, as it can affect your reaction time or whether you react at all. You must develop an offensive mind-set that will allow you to break through the fear barrier and defeat your attacker quickly and effectively. This starts by developing confidence in yourself, your training, and of course your fighting skills and techniques. You can know more than a hundred techniques, but if you do not have the proper mind-set for using them, they will be useless in a life-or-death situation.

Next, you must have a thorough knowledge of the vital, or vulnerable, points of the body and how the body reacts to the manipulation of these points. The human body reacts the same in most individuals when one of these points is struck; this is called an autonomic reflex. For example, if a man is struck in the groin, his hands will automatically grab for his groin to help ease the pain. Remember that for every action there is a reaction. By learning the autonomic reflexes associated with the various vital points of the body, you will gain an edge in any close-quarter combat situation. Knowing how your attacker's body will react to the first strike will enable you to set him up for multiple subsequent strikes.

THE FOUR LEVELS OF CONSCIOUSNESS

Unconsciously Incompetent

This is the level where you have no knowledge of the subject matter.

Consciously Incompetent

At this level, you are aware that a skill exists, but you have no realistic idea of how to use it.

Consciously Competent

This is the level at which you are aware of and practicing skills learned. You can execute the skills as long as you have time to think about them first.

Unconsciously Competent

This is the final level of mastery. At this level you don't have to think about reacting; your reaction to an attack is instantaneous and second nature.

NOTE: The final level—unconsciously competent—is the level you should strive to reach. When engaging in close-quarter combat, seconds can mean the difference between living and dying. You will not have the time to think about your response; you must react instinctively.

PRINCIPLES OF COMBAT PSYCHOLOGY

- **Strategy**—Your opponent's actions will dictate your next move. Never react out of anger; stay in control at all times.
- **Role Playing**—You must give your attacker the impression that you are vulnerable. This can cause the attacker to relax and become sloppy, thus setting him up for your counterstrikes.
- **Field of Vision**—Never look at or focus on any one part of your attacker's body. Use your peripheral vision to see everything at once; this will help prepare you for any strike coming from any direction.
- **Defusing the Fight**—Never antagonize or feed the fire. Never say or do anything that may bring the situation to a heightened level or put your attacker on the defensive (if you back him into a corner, you leave him no alternative but to fight).
- **Level of Force**—Legally speaking, you must meet equal force with equal force; however, this is not the case in combat directed by the U.S. government.
- **Fear**—Fear is your worst enemy. You must not allow fear to control your emotions or actions.
- **Anticipating**—Never anticipate what your attacker might do.
- **Natural Stances**—Always use natural stances and body movements so you don't telegraph your intentions.
- **Spontaneous Reaction**—Use your natural reflexes to enhance your movements.
- **The Four Ranges of Combat**—These are boxing, kicking, wrestling, and grappling.
- **Force Redirection**—Avoid meeting force with force. Always redirect your attacker's force and use it against him.
- **The Offensive Mind-Set**—This will allow you to strike first without hesitation and without fear of being injured by your attacker.
- **Body Language**—Never telegraph your intentions (physically or verbally) to your attacker.
- **The Four Levels of Consciousness**—These are unconsciously incompetent, consciously incompetent, consciously competent, and unconsciously competent.

Strategy

A good fighter is a strategic fighter, one who plans his moves carefully and can adapt to any fighting situation or attack. To be strategic is to be knowledgeable in the basic principles of close-quarter combat and to use these principles to your advantage.

TELEGRAPHING

Telegraphing is the process of showing your intent to make a certain offensive or defensive move before you execute that move. This surrenders your advantage. Avoid telegraphing unless you are using it as a ruse or stratagem to set your opponent up for your primary technique.

ANTICIPATING

When you anticipate or think something is about to happen, you tend to focus your thoughts on that specific action. In close-quarter combat this can quickly cause your defeat. For example, if you anticipate your opponent throwing a punch at you from his right side, you prepare your body to defend to your left. But happens if your opponent attacks from his left? Never try to anticipate your opponent's intentions, because close-quarter combat is often unpredictable.

FLOWING

To eliminate the possibility of anticipating your opponent's next move you must be able to *flow* with your opponent, making yourself one with him. To do this you must first be confident in your fighting skills and relaxed in the execution of your tactics. The use of your peripheral vision also helps by allowing you to see and take in the whole picture and all your opponent's moves, thus allowing you to flow with him.

DECEIVING

One of the oldest tricks in the book of boxing is to deceive your opponent by making him think you are hurt or that he is getting the best of you. The logic here is that if your opponent thinks he is getting the best of you, he will tend to become overconfident and sloppy in his technique, giving you the opportunity to counter his moves effectively.

PREMEDITATING

Never premeditate or plan a technique. This falls into the same realm as anticipating, but in a different sense. To premeditate your actions is to tell yourself, for example, *when this guy punches, I'm going to block his punch with my left hand and counterpunch him in the face*. Since you have just programmed your mind and body to take this action, that is what you will do. But what happens if your opponent doesn't punch, but kicks you in the groin instead? Get the point? Again, remember to flow with your attacker.

FORCE REDIRECTING

It is often easier to redirect your opponent's force than to meet it head on. The art of aikido is based on this concept, and it is used quite effectively by those who have mastered its principles. Force redirection is simply the process of redirecting your opponent's force (punch, kick, or push) to a position of advantage to you and disadvantage to him. This can be accomplished, for instance, by getting your body out of the line of attack and allowing your opponent's momentum to continue in the same direction, thus making him vulnerable to a throw, takedown, or strike.

Defense against Punches

The techniques covered in this chapter are designed to incapacitate, immobilize, or kill your attacker, if necessary, with a rapid progression of moves from defense to offense. Therefore, pay attention to detail, and use care and control while practicing to prevent injury to you or your partner.

For the purpose of clarity regarding the mechanics of each technique, the demonstrators are wearing plain BDUs without extra combat gear.

TECHNIQUE NUMBER ONE

As the attacker steps in to punch, execute a left-hand outside block while stepping to the right (getting your body out of the line of attack).

Execute a right-hand palm-heel strike to the attacker's chin, throwing him off balance.

Step through with your right leg and sweep the attacker's leg out from under him.

Once the attacker is on the ground, grab his arm just above the elbow and turn him onto his stomach, then place him in an armlock for control.

TECHNIQUE NUMBER TWO

As the attacker begins his punch

execute a left-hand outside block.

Now counterstrike with a right-hand punch to the attacker's throat.

Continue with an uppercut to the attacker's ribcage

followed by grabbing the attacker's groin with your right hand.

Finish up by pulling the attacker's groin outward while pushing him backward with your left arm, taking him to the ground. Follow up.

TECHNIQUE NUMBER THREE

As the attacker steps in to punch,

execute a left-hand outside block.

Immediately step in with a right elbow strike to the attacker's jaw, keeping the momentum of the strike going.

Below left: Spin around with a second elbow strike, your left elbow

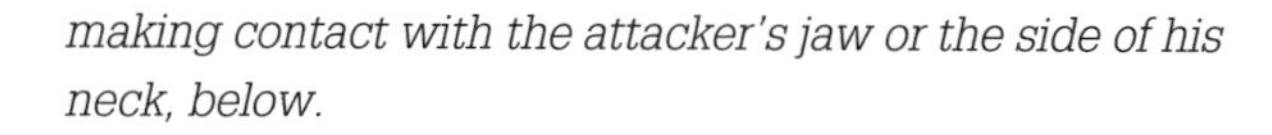

making contact with the attacker's jaw or the side of his neck, below.

Offensive Knife-Fighting Tactics

A knife in the hands of a skilled fighter can be lethal and extremely hard to defend against. Technically speaking, it is often easier to defend against a gun at close range—where the trajectory of the bullet is in a straight-line path—than it is against a knife coming at you from various angles, one right after the other.

This chapter will address the two basic grips and offensive knife-fighting drills/techniques.

A WORD ON CRITICAL TARGET POINTS

Before we get into specific techniques and their respective target points, we should examine some especially critical target points

REVERSE GRIP

This is the most preferred grip. Hold the knife so that the blade is parallel with your forearm, making sure that the blade is facing outward. There are two major advantages to using the reverse grip: first, the knife can be easily concealed under the forearm and out of view until you are ready to strike, and second, because of the position of the knife it is harder for it to be knocked out of your hand.

STRAIGHT GRIP

Lay the knife handle diagonally across your palm and close your hand, gripping the handle with the blade pointed toward your attacker, the edge of the blade facing downward. This grip can be used to jab, stab, or make quick slashing motions to inflict numerous cuts. When using this grip, keep your knife hand back until you are ready to strike, to prevent your opponent from kicking or knocking the knife out of your hands.

and what results to expect when you attack them with a knife.

Forehead

A slash across the forehead can cause blood to flow into your attacker's eyes, blurring his vision.

Eyes

A knife strike directly into the eye socket can easily blind the attacker. The eye is also an excellent route to the brain.

Mouth

A deep stab into the mouth can cause heavy bleeding in the throat, which can choke the attacker. A deep enough stab can also damage the brain stem.

Jugular Vein and Carotid Arteries

Any strike delivered forcefully to the throat or neck can be lethal if the jugular vein or one of the carotid arteries is severed. The jugular is a major vessel carrying blood from the head back to the heart, and the carotid arteries carry oxygenated blood away from the heart.

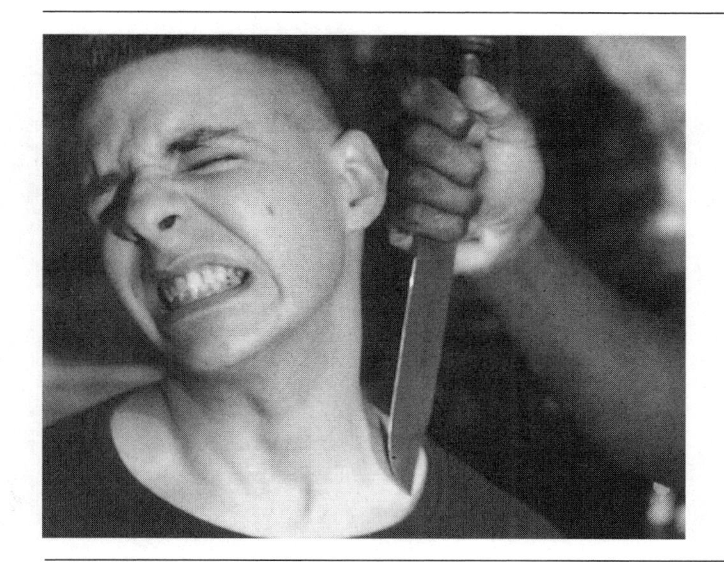

Subclavian Artery

This artery lies just below the collarbone (clavicle). A cut here will cause immediate internal bleeding followed by death.

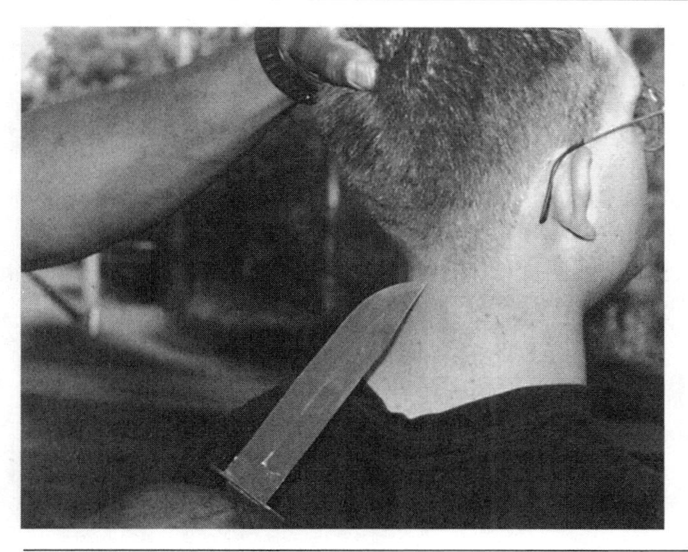

Medulla Oblongata (Brain Stem)

This is the lowermost portion of the brain, located just above the uppermost cervical vertebra (the base of the skull). When you attack vigorously with a knife, you can sever the medulla oblongata, causing instant immobilization and death by cutting off all electrical impulses from the brain to the rest of the body.

Thumb-Forefinger Web

An excellent distraction technique is to cut the web between the thumb and forefinger. This can greatly affect your attacker's ability to grasp.

Inside of Wrist

A slash to this area can cause excessive bleeding and aid in distracting your opponent.

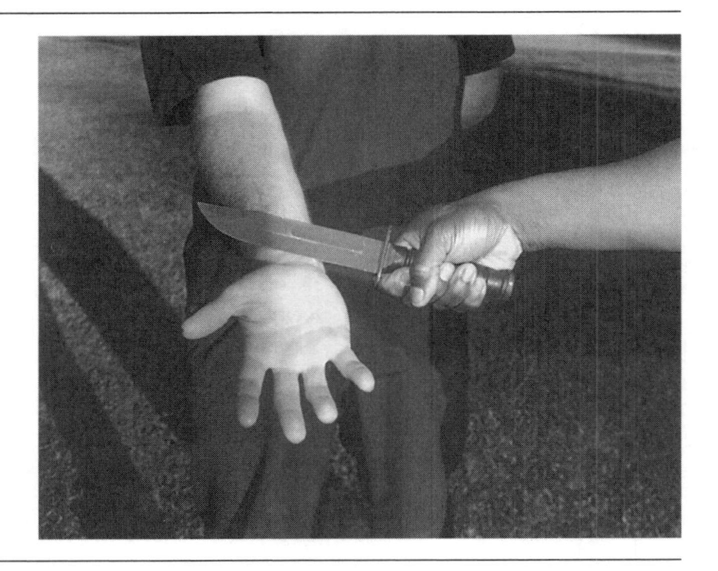

Armpit

A knife strike deep into the armpit can sever major nerves and immobilize the entire arm.

Heart and Lungs

The heart and lungs are located just behind the ribcage. By turning the knife so that the blade is horizontal you can easily slip the ribs and puncture the heart or lungs, causing instant or near instant incapacitation and death.

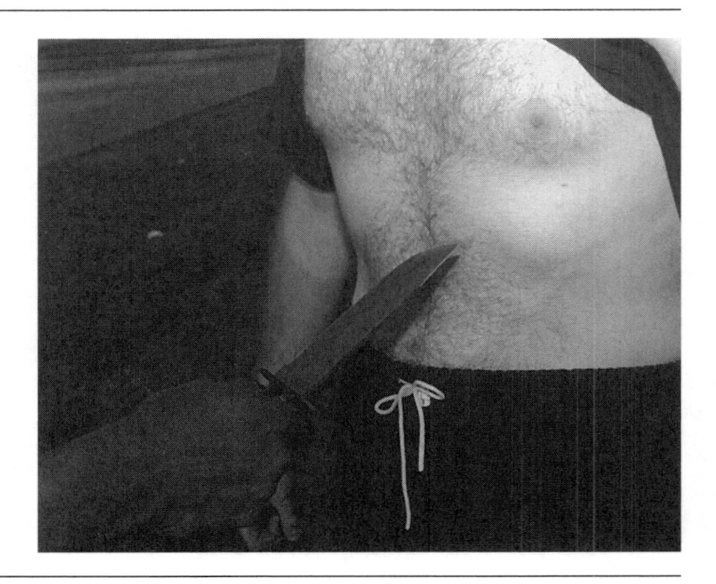

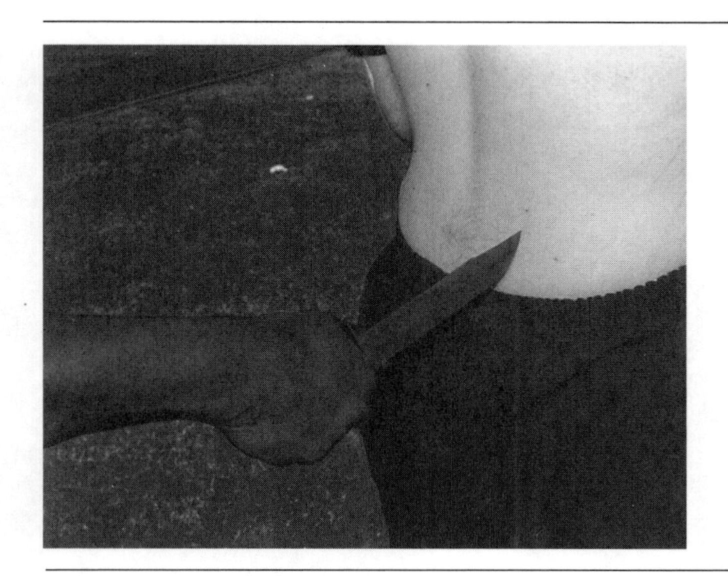

Kidneys

Considered a "high shock" target, the kidneys are open to stabs and slashes. When this happens, death is almost certain.

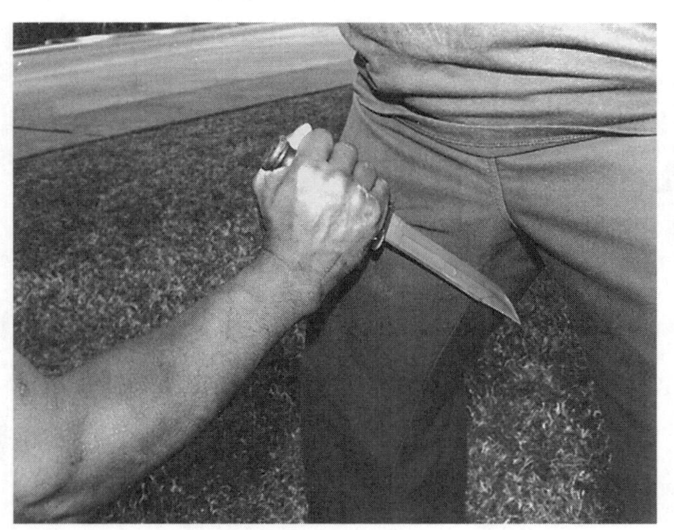

Groin

Though the groin is not always considered a lethal target, the psychological effect produced by a knife strike there can be devastating, drawing your opponent's attention to the wound and allowing you to follow up with another strike.

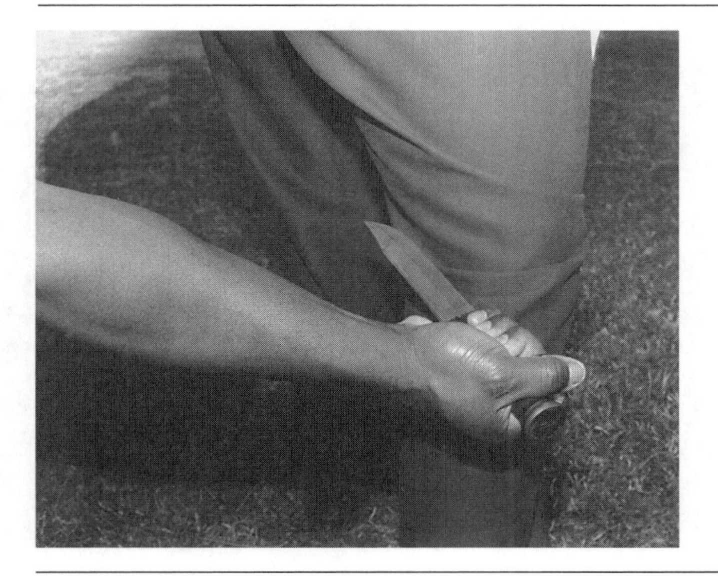

Back of the Knee

A knife slash to the back of the knee can easily sever major muscles and tendons used for mobility.

BASIC OFFENSIVE KNIFE-FIGHTING TECHNIQUES

Target Points

Target points include the eyes, mouth, throat, neck, chest, solar plexus, ribcage, midsection, and groin.

Straight Thrust

Hold the knife in the straight grip and thrust it straight into the target.

Target Points

The throat, sides of the neck, inside of the wrists, backs of the knees, and groin.

Forward Slash

The forward slash can be executed using the straight or reverse grip and is delivered by bringing the knife across your front.

Target Points

The throat, sides of the neck, backs of the knees, groin, and inner thighs.

Reverse Slash

The reverse slash, like the forward slash, can be employed with either the straight or reverse grip but is delivered in the opposite direction.

Defense Against Knife Attacks

Effective knife defense starts with a proper mind-set and attitude. Too many instructors teach the physical side of knife defense and neglect to teach the psychological aspects as well the results that are likely to occur if a student is ill-prepared to handle a knife fight.

Chapter 9 (Combat Psychologyand Mental Conditioning) covers the development of the proper mind-set in detail. However, regarding knife defense, it should be noted that your mind-set should be, "I'm going to get cut, but I'm not going to get killed." The reason for this mind-set is simple: by mentally preparing yourself for injury you will be less likely to go into shock in the event that you are cut. If you take even a second to take your eyes and attention off of your attacker to assess a cut, your attacker will take advantage of this opportunity and deliver a lethal strike.

The techniques covered in this chapter address defense against knife attacks from various angles. When training with your partner, you should use a rubber knife as a training aid to lessen the chance of injury. However, a wooden knife offers more realism in training because of the feel of the wooden blade upon your body when contact is made.

OVERHEAD ATTACK

As the attacker steps in with an overhead attack, execute a high block while keeping your body low and poised for your counterattack.

Execute a palm-heel strike directly to the attacker's chin, snapping his head back.

Wrap your right arm around the top of the attacker's arm and grab your biceps, placing him in a figure-four armlock.

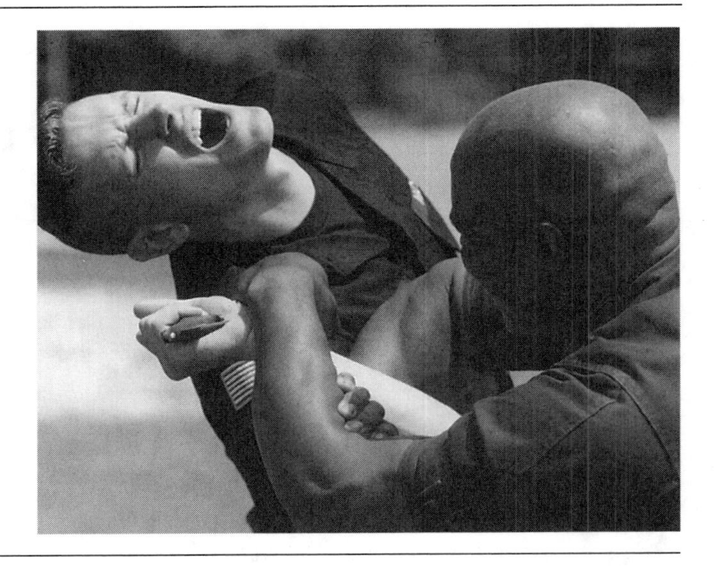

Step through with your right leg as you continue to bend the attacker's body downward.

Execute a rear-leg takedown while controlling your attacker's knife hand. Disarm the attacker and follow up.

STRAIGHT THRUST

As the attacker steps in with a straight thrust, parry your body out of the line of attack and block the knife hand away from you.

Immediately wrap your left arm around the top of the attacker's knife hand, while maintaining control of his wrist with your right hand.

Disarm your attacker by bending his wrist sharply downward.

Follow up with a forearm strike to the throat.

Take the attacker to the ground with a rear leg sweep and

follow up.

UPWARD THRUST

As the attacker steps in for an upward thrust, execute a low cross block (right over left in this instance).

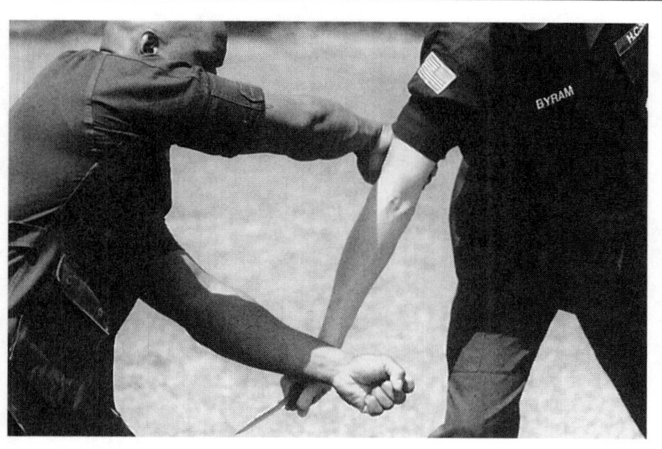

Next, slide your right hand up the attacker's arm to just above his elbow joint.

Below left: Pull the attacker's arm at the elbow, causing his arm to bend, while bringing your left arm behind the attacker's shoulder, placing him in an armlock.

Below right: Maintaining the armlock (control of the knife hand), force the attacker to the ground and follow up.

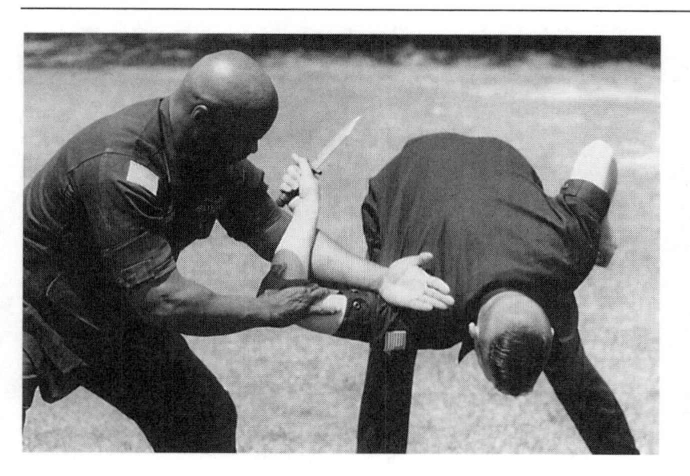

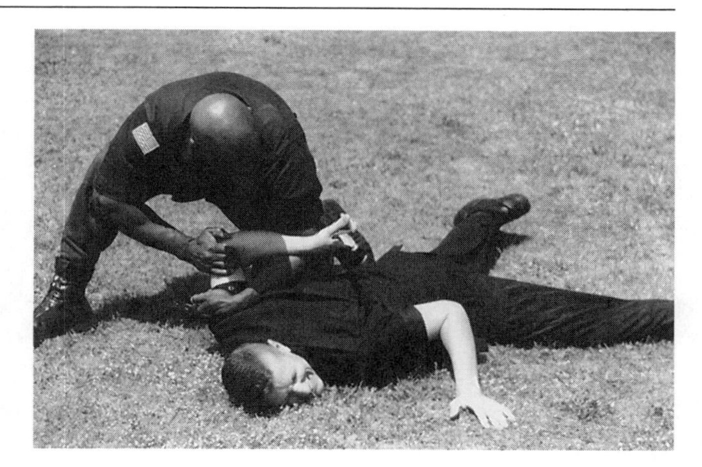

Choking Techniques

Chokes are used to disrupt oxygen (blood flow) to the brain, causing the attacker to lose consciousness within a matter of seconds. Chokes are highly effective "kill" techniques, and extreme caution and care should be exercised when practicing the choking techniques covered here. When you are applying these chokes to your partner, he should use the "tap out" signal by patting your leg or side to indicate that he feels the effectiveness of the choke.

CAROTID CHOKE

This is applied to the external carotid arteries, which are located on the left and right side of the trachea. The choke is applied by pinching off the arteries, stopping blood flow to the brain. When this is applied correctly, unconsciousness can result in five seconds.

CAROTID SLEEPER CHOKE

This choke is applied by bringing your left forearm around your attacker's throat from behind.

Grab your right bicep with your left hand and then

bring your right hand up and behind the attacker's head.

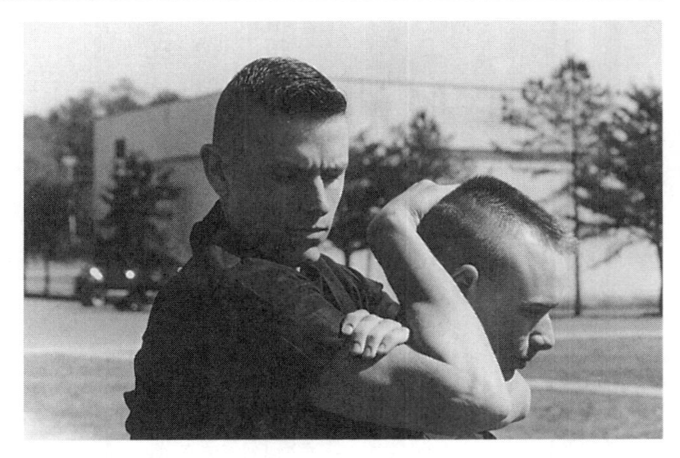

Once you are in this position, apply the choke using rearward pressure on his throat with your left forearm and forward pressure on the back of his head with your right hand.

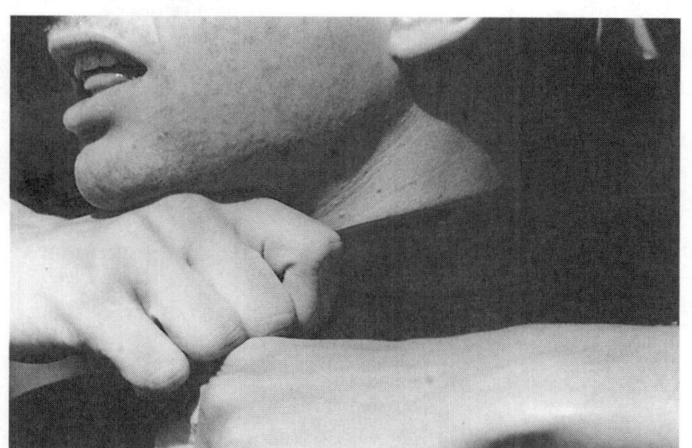

CROSS-COLLAR CHOKE

Top and left: This choke can be applied from either the front or the rear by crossing your wrists and grabbing your attacker's collar, then pulling the opposite sides of the collar outward.

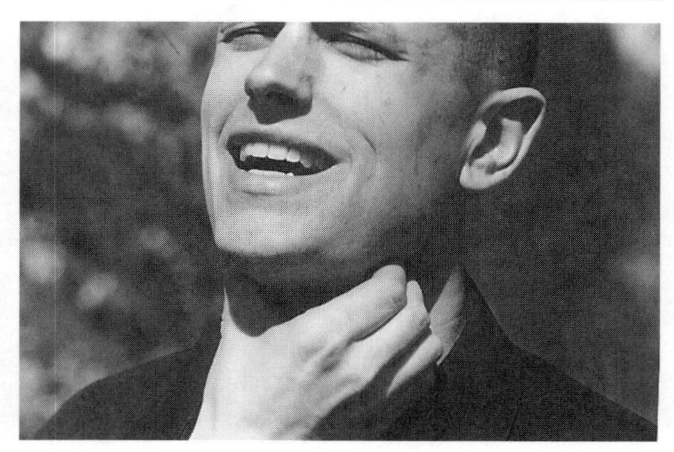

TRACHEA CHOKE

Lower left and above: This deadly choke is applied by grabbing the attacker's trachea and squeezing. If killing the attacker is required, simply crush the trachea.

COLLAR CHOKE

Like the cross-collar choke, this choke is applied by grabbing the attacker's collar and pushing inward into the trachea.

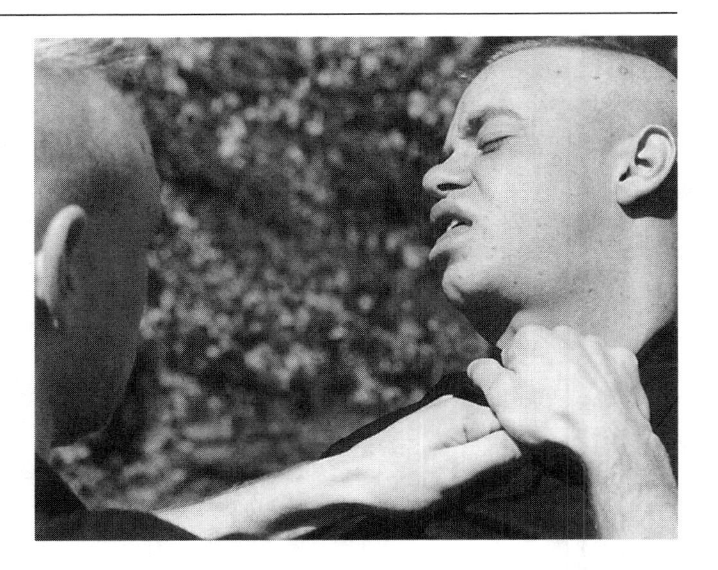

Breakfalls

This chapter will cover the four basic breakfalls used in close-quarter combat training. It is important that you pay close attention to detail regarding proper body positioning, including head, leg, and arm placement. Emphasis will also be placed on the proper expulsion of air from the body upon making contact with the ground. You should perfect all breakfalls covered here before moving on to throwing techniques.

LEFT SIDE BREAKFALL

Lying on your left side, hold your left leg almost straight, with a slight bend at the knee.

Your right knee is bent and your foot is flat.

Your left arm is at a 45-degree angle to your body, and your right hand is protecting your upper torso.

Raise your head off the ground by tucking your chin in; this will help to avoid a head injury. Execute the breakfall by rolling to your right and assuming the same breakfall position.

RIGHT-SIDE BREAKFALL

The right-side breakfall is executed in the same manner as the left-side breakfall.

FRONT BREAKFALL

Start from the kneeling position.

Raise both of your arms so that your hands are at ear level.

Allow your body to fall naturally forward.

Break your fall by making contact with the ground with your hands and forearms while turning your head to the side to avoid facial injury. Yell loudly upon making contact with the ground; this helps you absorb the impact of the fall.

REAR BREAKFALL

Start from a sitting position with your arms folded across your chest and your knees bent to about midchest.

Below: Execute the rear breakfall by rolling naturally backward, keeping your chin tucked in and your head off of the ground to avoid injury.

Below right: Slap the ground with your arms at the same time your back makes contact with the ground; this lessens the impact of the fall.

Throwing Techniques

It is important to remember that the key to the execution of a successful throw is technique and not mere strength. Although physical strength does play a vital role in this process, proper technique includes the setup or off-balancing of the attacker and entry and execution of the throw.

Your attacker must first be put off balance. The setup is done by moving or redirecting your attacker's momentum in the direction of the intended throw.

Without proper entry going into the throw, the chances of self-inflicted injury as well as being countered are enormous. Therefore, proper entry is crucial for successful execution of your throw.

Following are some of the most devastating throws used in close-quarter combat. Do not practice these throwing techniques until you and your partner are fully versed in the breakfalling drills presented in the previous chapter.

ONE-ARM SHOULDER THROW

As the attacker steps in to punch, block his arm with an outside block and step in with your right foot toward his lead foot (this will allow you to pivot when throwing the attacker). Hook your right arm under the attacker's punching arm.

Pivot your rear leg around, bringing both of your feet together (having them spread apart can injure your lower back when you attempt to throw). Ensure that you get your hip over enough so that the attacker is on your back and not sliding off your side.

Below: Raise the attacker off the ground,

throw him to the ground, and follow up, below right.

HEAD THROW

Grab the attacker around the head and position your body as with the one-arm shoulder throw.

Below left: Now, instead of throwing him over your back, stick your leg out past his foot.

Below: Turn your body and throw him to the ground.

ELBOW-TO-SHOULDER THROW

Grab the attacker, and as you step in to throw,

shoot your elbow underneath his punching arm while punching him under the chin.

Continue your movement until you are in a stable throwing position, then

execute the throw.

REAR LEG TAKEDOWN

Block the attacker's punch and

step into him (below), breaking his balance to the rear (this is important in keeping him from using the same technique on you).

Below right: Now execute a leg sweep.

FLYING SCISSORS TAKEDOWN

While squaring off with the attacker,

quickly cross-step while keeping your body toward him and your eyes on him.

Quickly throw your body into the air and wrap your legs around his legs, twisting your body in the direction you want him to go.

Defense against Holds

This chapter will cover defenses against the most common holds you are likely to encounter on the street or on the battlefield. Emphasis is placed on the use of distraction tactics to draw the attacker's attention away from the applied grab, thus opening him up for your counterattack.

FRONT LAPEL GRAB

As the attacker grabs you,

reach up and grab his right wrist with your left hand.

Strike the attacker with a hook punch to his jaw and follow through far enough to set him up for a

quickly executed reverse elbow strike

to the side of the neck.

REAR BEAR HUG

As the attacker grabs you from behind,

distract him by striking him on the bridge of his nose with a rear head butt,

by grabbing his groin and crushing his testicles,

or by stomping on his instep (your heel should make direct contact with his instep and not his toes).

Step to your left with your left foot, thus placing your attacker's left foot between your legs (his stance will usually be wide for stability as he grabs you).

Reach down between your legs and grab him by an ankle, forcing your buttocks into his midsection.

Pull his leg straight up, throwing him off of you while maintaining control of his ankle.

Once the attacker is on his back, sit on his knee while maintaining your grip on his ankle.

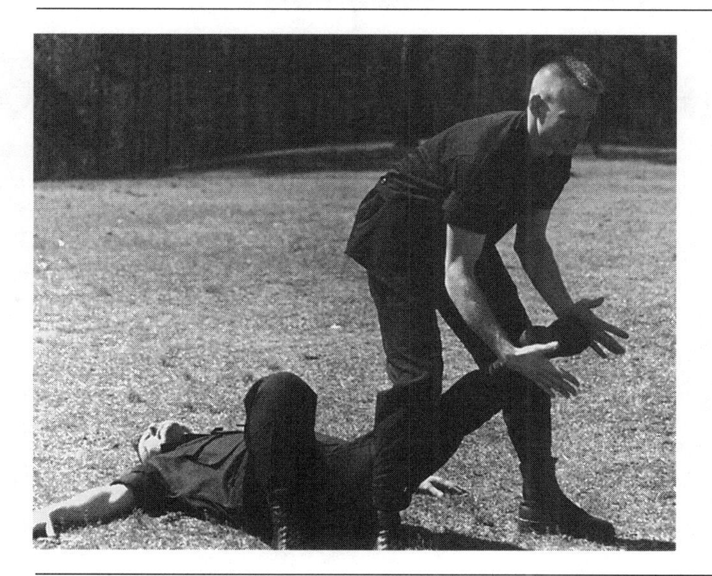

Twist his ankle sharply and break it at the joint. You can also execute a stomp to his groin with your right foot.

FRONT BEAR HUG

As the attacker grabs you,

execute a head butt to the bridge of his nose.

Reach between his legs, crush his testicles, and push him away from you.

REAR HEADLOCK

As the attacker grabs you, keep direct contact away from your throat by turning your head to one side.

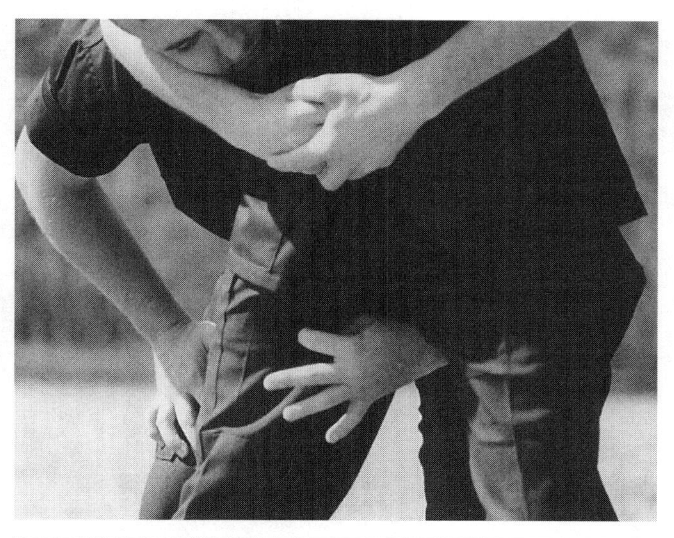

Reach between the attacker's legs from behind and pinch the nerve point approximately two inches below the upper inner portion of his thigh, or just crush his testicles.

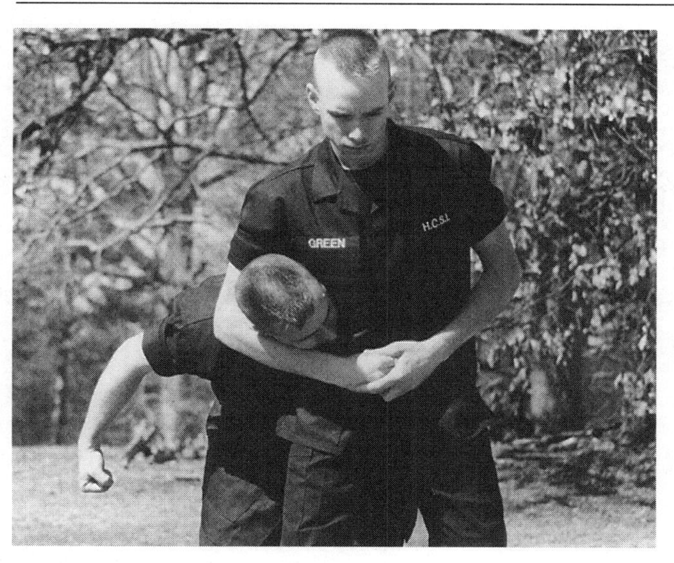

Strike the back of his knee with a hook punch,

breaking him down.

Reach up behind him and apply force beneath his nose with your middle finger.

Below: This is an effective pressure point when applied correctly; notice the position of the middle finger.

Below right: Finish by pushing upward under the nose and taking the attacker off of you.

FULL NELSON

As the attacker grabs you,

reach up with both hands and grab his fingers, then peel them away from your head at the knuckles.

Below left: Grab his right hand and pull it sharply away from your head (this technique also keeps the attacker's hand away from your sidearm, should you be wearing one).

Below: Quickly turn to your left and place your right hand under his left elbow.

Keeping control of his fingers with your left hand (notice the rearward bend of the attacker's finger) and control of his elbow with your right hand,

manipulate the arm by bending it backward,

taking him to the ground.

Defense Against Handguns

If you are in a situation that requires disarming your attacker of a pistol, the same mind-set applies as if you were defending yourself against a knife attack: "I may get shot, but I won't get killed." This chapter will present several scenarios and illustrate the best defense to use for each.

Remember, the most important factor in disarming an attacker who has a handgun is to get your body out of the line of attack, the line of attack being the trajectory of the bullet. This can be accomplished by parrying your body to the left or the right while pushing the weapon off of your centerline and away from any vital organs. You may catch some flash burn from the muzzle, or the bullet may graze your skin (a flesh wound), but you can still fight and defend yourself as long as your vital organs are not injured.

The use of distraction or deception tactics is also important. By diverting the attacker's attention away from the immediate situation, you can provide yourself with a window of opportunity to disarm him.

NOTE: Do not attempt to disarm your attacker if he is out of your reach (the bullet will reach you faster than you can reach him). Once you engage your attacker you must follow through completely without hesitation to disarm and neutralize him quickly.

As the attacker places the muzzle to your chest, raise your hands to ear level. You should do this for two reasons: first, it tells your attacker that you are unarmed and will not try anything, and second, it places your hands in a position of advantage closer to the weapon.

PISTOL TO CHEST

Quickly attack the wrist of the gun hand, bending it sharply and

causing the hand to open, thus disarming the attacker.

Keeping control of the wrist, apply whatever pressure is required to gain pain compliance from the attacker. Take him down and follow up by breaking his wrist.

PISTOL TO BACK OF HEAD

When the attacker places a muzzle to the back of your head, you will have a good idea where his arm is.

Quickly turn to one side while directing the weapon away from you, keeping out of the line of attack.

Immediately strike the attacker's eyes, blinding him. Follow up immediately with a finger strike to the eyes, followed by a joint lock to the arm and then a palm-heel strike to the underside of the chin or a punch to the throat.

PISTOL TO SIDE OF HEAD

As the attacker grabs you from behind with the muzzle to the side of your head,

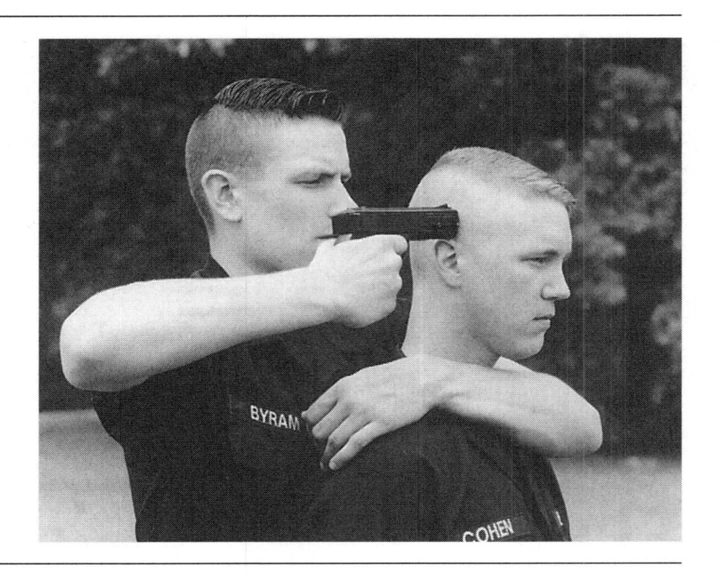

quickly turn your body to the left while pushing the barrel away from your head and simultaneously striking him in the groin.

Below: Bring his gun arm down and away from your body while hooking your right arm underneath his gun arm, placing him in an armlock.

Below right: Take him to the ground, breaking his arm.

PISTOL TO UPPER BACK

As the attacker places the muzzle in your back, keep your hands raised and slightly turn your head to see the attacker's position.

Quickly turn your body out of the line of attack, pushing the gun hand farther away from your body.

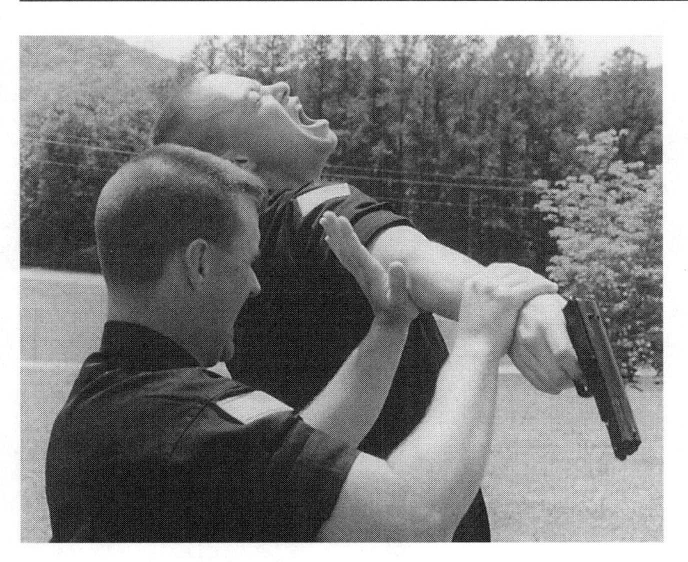

Grab the gun hand and place your opposite hand at the back of the attacker's elbow.

Apply direct pressure to the elbow,

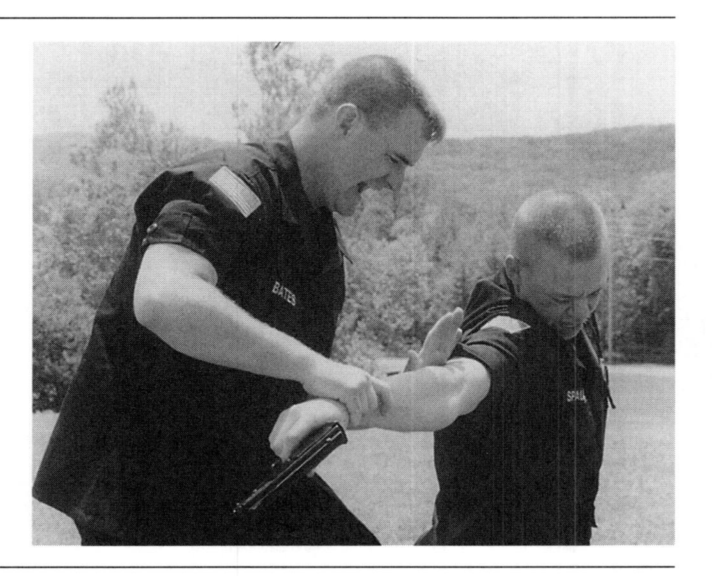

taking the attacker to the ground. Follow up by breaking his arm.

Ground-Fighting Techniques

There is an excellent chance that a fight that starts from an upright position will go to the ground. In this case, ground-fighting tactics must be employed. The main objective in a ground fight is to gain the advantage over your attacker as quickly as possible.

The following techniques will cover effective pins, escapes, and counterstrikes used in ground fighting.

WRIST PIN

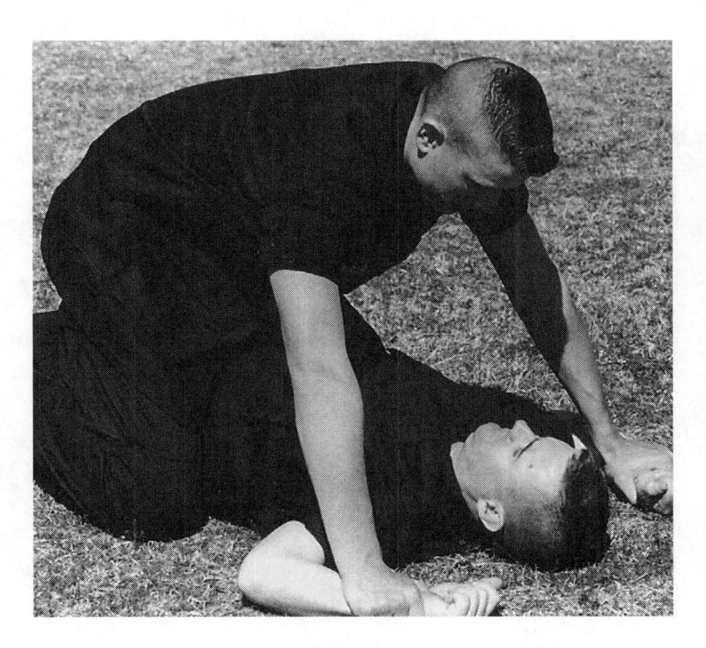

The attacker has you pinned on your back by the wrists.

Stabilize yourself by bending one knee and executing a strike with the other knee to the attacker's coccyx. At the same time, bring both of your hands down to your sides, breaking your attacker's balance and causing him to fall forward. (Turn your head to one side to avoid chest-to-face contact.)

Grab him around the waist and

roll him over to

place yourself in the position of advantage, below.

Follow up with a strike to his groin, below right.

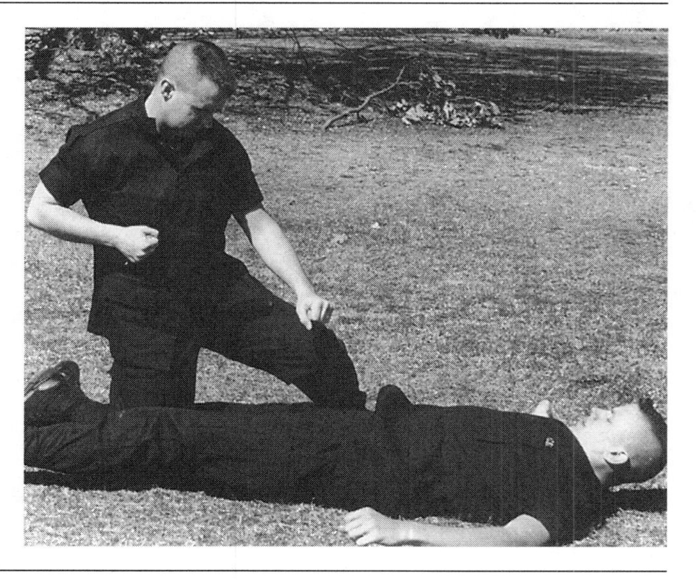

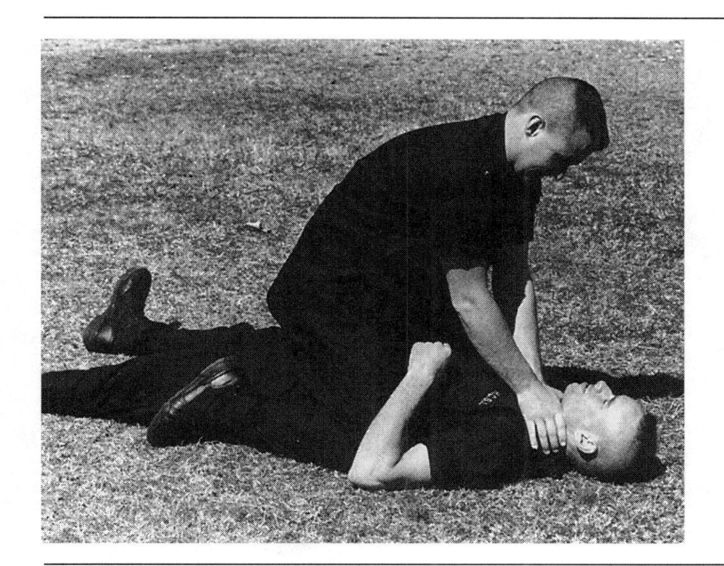

FRONT SUPINE CHOKE

The attacker has you pinned on your back, applying a choke.

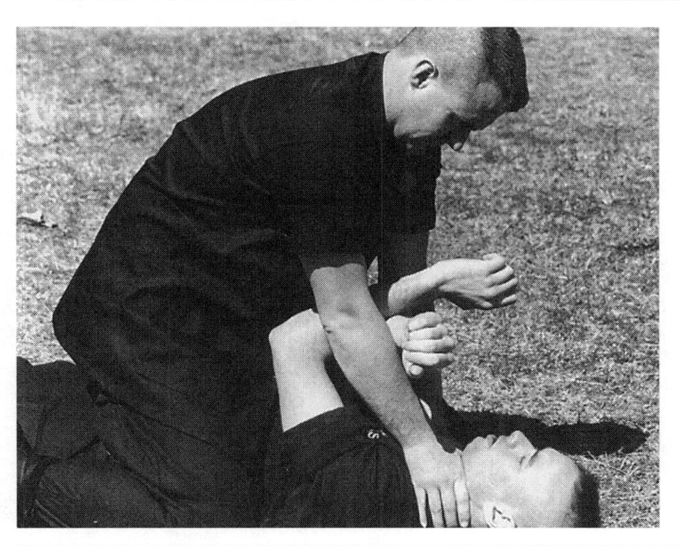

Quickly bring your arms up and between the attacker's arms.

Below left: Continue this motion to break his choke hold.

Below: Wrap your arms around his, breaking his balance and driving his head/face into the ground.

Keeping his arms pinned, roll him over

onto his back.

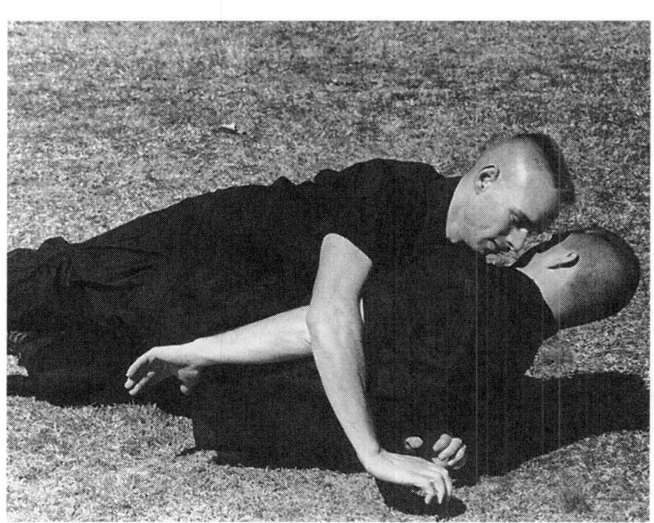

Finish up with a strike directly into the groin.

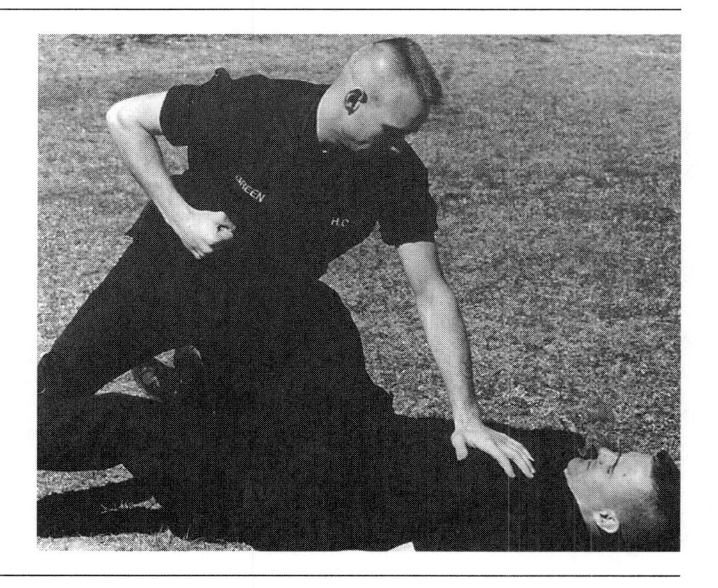

REAR PRONE CHOKE

The attacker has you pinned down on your stomach and is choking you with his hands around your neck (fingers against your trachea).

Reach behind your head and grab both of his wrists.

Break him down by pulling his wrists outward.

Raise your body and pull his right wrist forward hard to

throw him over your right shoulder and onto his back.

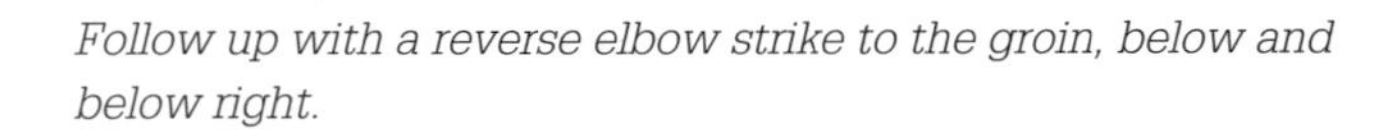

Follow up with a reverse elbow strike to the groin, below and below right.

Rifle/Bayonet Tactics

A soldier's primary weapon and first line of defense is often his M16A2 rifle. When coupled with a bayonet, the rifle becomes a formidable weapon for close-quarter fighting. This chapter will cover both defenses against rifle/bayonet attacks and offensive rifle/bayonet techniques.

BUTT STROKE

A butt stroke to the attacker's solar plexus or ribs is an excellent offensive tactic, but it should be followed up with the bayonet.

The butt stroke can also be directed in a downward motion, striking the back of the attacker's head or neck, or powerfully striking the side of the neck to knock him unconscious or break his neck.

SLASH

A slash to the side of the neck is lethal, severing the external carotid artery or the jugular vein and causing the attacker to bleed to death rapidly.

A slash can also be directed against the back of the knee, severing the muscles and tendons and disabling the leg.

THRUST

The thrust can be delivered against all target areas of the body. When directed into the attacker's solar plexus or abdomen or between his ribs ("slipping the ribs"), it becomes a lethal strike, piercing major organs and causing severe internal bleeding.

BLOCKS

Left and below left: The rifle can also be used to block punches and overhead attacks. Kicks can be blocked, too, while inflicting severe damage to the attacker's leg.

UPPERCUT BUTT STROKE

Below: This modified butt stroke, when directed against the chin, can easily dislocate or break the jaw; it is an excellent knockout shot.

Expedient Weapons

The average soldier carries enough field gear to survive the most extreme weather conditions in combat. This same gear can be used to defend oneself from attack.

This chapter will examine the most practical pieces of gear that can be used as effective weapons against an attack.

ENTRENCHING TOOL (E-TOOL)

The E-tool can be a devastating weapon when used properly. The blade of the E-tool can be adjusted to a straight or 90-degree angle, which can aid in certain blocking and striking maneuvers.

TENT PEGS

A tent peg can be used to jab, stab, or choke your attacker.

TENT STAKES

A tent stake can be used as a stabbing weapon. Target areas include the eyes, throat, and upper torso.

TENT ROPE

A tent rope can be used to strangle your attacker or tie him up and secure him.

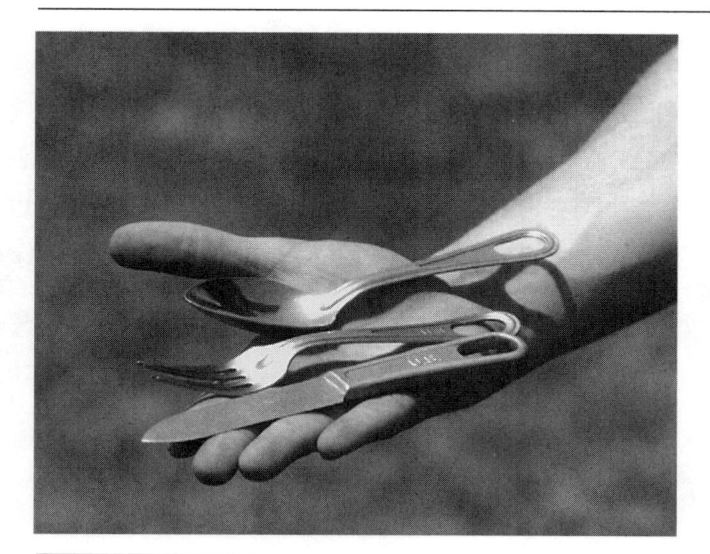

MESS KIT UTENSILS

The mess kit contains several weapons that can be used with lethal results. The knife, fork, and spoon can all be used as weapons for self-defense. The spoon's handle can be used to attack the brain via the eye.

LOAD-BEARING EQUIPMENT (LBE) OR "WEB GEAR"

Below: You can use your load-bearing equipment as a distraction against a knife attack by throwing it into your attacker's face or entangling him in the gear itself.

A Final Word about Safety and Training

THE SAFETY FACTOR

When conducting mass CQC training sessions, a safety briefing should be given before each phase of training. To avoid injury the instructor should also ensure there is approximately eight to 10 square feet of space between each pair of trainees. All jewelry, including rings, watches, necklaces, dog tags, and so on, should be removed. An adequate number of assistant instructors should be used to monitor all trainees at all times. Medical personnel should be on hand in the event that any injuries are sustained by trainees.

THE INSTRUCTOR

The primary or chief instructor should be highly competent and skilled in the art of hand-to-hand combat. The chief instructor should also be in top physical condition, capable of performing all exercises and training drills being taught. He must also be fully versed in his presentation and demonstration of all techniques and be ready to answer clearly and correctly all questions that trainees may pose during the training session.

ASSISTANT INSTRUCTORS

You and your assistant instructors (AIs) must be "on the same sheet of music." This means that you must work closely together to ensure that the techniques being taught are fully understood by the trainees. If there is a communication gap between the primary instructor and his AIs, different versions of a technique may be taught; this can lead to immediate confusion among the trainees. Your assistant instructors should also be fully versed in all presentation and demonstration techniques. An assistant instructor with prior martial arts experience can be an added bonus; experience must always be considered when selecting your training staff.

UNIFORM WEAR

Your trainees should wear what they are most likely to be wearing in a close-quarters combat situation. For the soldier, full combat gear is advised, to include Kevlar helmet, LBE (load-bearing equipment), BDUs (battle dress uniform), and leather boots. This uniform can be modified depending upon the type of training being taught at the time. For example, when you are conducting GFT (ground-fighting tactics), the LBE and Kevlar helmet should be removed for safety purposes; however, full combat gear should be worn during stationary drills (striking and kicking). Likewise, in training civilians should wear what they are most likely to be wearing if attacked on the street, at home, or at work.

Remember, you must train as you intend to fight if you expect to survive and triumph on the battlefield.

TRAIN TO FIGHT, FIGHT TO WIN!

The difference between victory and defeat in a close-quarter combat situation is knowledge of your opponent's physical and psychological weaknesses and how to exploit them to your advantage. The information contained in this book will give you the knowledge to emerge victorious in close-quarter combat.

Remember, *knowledge is power*. You now have it!

Train to fight, fight to win!

About the Author

Professor Holifield has been involved in the martial arts for more than 30 years and has served as a subject matter expert, trainer, writer, and consultant in the field of hand-to-hand combat and unarmed self-defense for the U.S. Army and various law enforcement agencies. His work has been featured in several major U.S. Army publications, including *Soldiers*, *Army Trainer*, and *Infantry*, and on the Armed Forces Korean News Network.

A 8th degree black belt and member of the World Martial Arts Hall of Fame, Professor Holifield is a world-certified instructor and has trained hundreds of soldiers army-wide. During Operation Desert Storm he was credited with capturing one of the first Iraqi soldiers. His worldwide organization, Holifield's Combat Systems International (H.C.S.I.) continues to draw members from across the globe who seek a no-nonsense, real-world approach to unarmed self-defense. This has made him one of the most sought-after instructors in the country.

Professor Leonard Holifield.